Natural History for Young Folks:

What Lizards Do

Cover Photo:
Crested Basilisk on a Twig

Natural History for Young Folks:
What Lizards Do

Brown Iguana

Fenton R. Kay, Ph.D.

KayLibros KL Press

Dedication

This book is dedicated to all the people in my long career as a biologist who have shown me the value of early understanding of the natural history of animals.

Table of Contents

List of Lizard Pictures

Why Did I Write This Book?

As a very young person, I was curious about everything in nature and constantly looked for books about the animals I saw in the desert where I lived. This is the kind of book I would have loved back then. Because of my early reading, I became a herpetologist—a lizard-and-snake nut—in college and helped start an amateur herpetology group. As a graduate student, I spent a year in the laboratory of one of the pioneers of lizard behavior. This book represents my efforts to present the natural history and behavior of lizards to young readers like you in a way I would have enjoyed.

I hope this small book will brighten your eyes and help you discover nature.

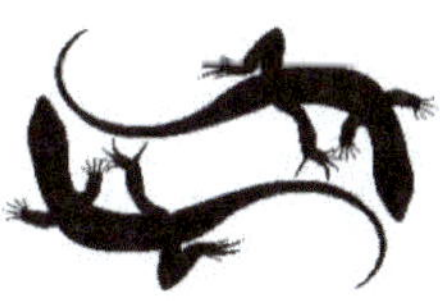

What Lizards Do

Staying Warm and Cooling Off

Collared Lizard Displaying and Sunning on a Rock

So, how do you stay warm on a cold day? How about staying cool in the hot summertime? I'll bet you have a nice heater and a fan in your room or central heating and cooling in your house. Well, guess what? Lizards

Wall Lizard Sunning on Rock.

don't have those things. If you put a lizard in a cool place, it will become cool. If you put one in a warm place, it will become warm. People call animals like that cold-blooded. Cold-blooded is not really a good word to describe what lizards do. Animals like us that always have a warm body temperature are called endotherms – internal heaters. Critters like most fish, frogs, toads, salamanders, snakes, turtles, and lizards are called ectotherms – outside heaters. Lizards come in two kinds when you talk about heating and cooling. Some are called heliotherms—basically, the word means sun warmers. The others are called thigmotherms, which means ground warmers.

Sun warmers get warm by sitting in the sunlight and warming up.

Side-blotched Lizard Sunning on Red Sandstone Rock with Lichens

You know how that works. You get out of the swimming pool and are chilly, so you find a nice recliner in a sunny spot, lie down, and warm up—aha, you are a heliotherm. OK, so how does a sun warmer cool off? You would jump back into the pool, and some lizards do that, too. Most sun warmers just get out of the sun—they find a nice shady place that is cooler and stay there for a while. Isn't that a neat way to get warm and stay cool?

But how about in the winter when it is cold, even in the sunshine, or it is snowy and cloudy? Sun warmers go to sleep for the winter, or most of it. Some lizards live in places that are warm all year, like tropical forests or where the winters are not too cold. Tropical sun-warming lizards bask in the sun in the early morning when it is cool, but can spend most of their days looking for food or a mate. Some small sun-warming lizards that live where winters are cold but not too cold will

Immature Five-lined Skink Warming on Wooden Steps

come out and sit on rocks in the sun on warm days. They can do that

Western Fence Lizard Warming on a Rock after a Cool Night

because they are small and warm up easily. Big sun-warmers spend the cold time asleep under the ground.

Now, the ground warmers do things a little bit differently. They don't sit in the sun; they hug the ground, rocks, or anything warm. When they get enough heat from whatever they are on, they go about the business of finding food or mates, just like the sun warmers. Most ground warmers live in places, like tropical forests, that never get very cold. It is always warm in those places so that the ground warmers can do their thing all day and sometimes all night.

That fence lizard bobbing its head at you on top of your wall is a sun warmer. If you go out early in the morning, you might find your fence lizard hanging on the sunny side of the wall, getting warm. The little gecko that climbs up your window screen at night is a ground warmer.

Blue-tailed Skink Warming on a Log

Ground warmers, such as geckos, can live in people's houses because houses are nearly always warm, and geckos don't need sunshine to stay warm.

Mediterranean House Gecko Warming on a Wall

Fenton R. Kay

Push-ups and Head-bobs

You have just left your summer cabin in the mountains to enjoy the early morning sun. On a tree stump nearby, you see something moving.

Male Western Fence Lizard Doing a Pushup on a Stump

"Hey, what's that? A lizard doing push-ups on that tree stump - up, down, up, down--what's going on?"

The lizard stops doing push-ups and hugs the tree stump.

What Lizards Do

"Well, OK, the lizard likes the morning sun," you mumble, "but doing push-ups?"

So, you sit on the porch steps in the warm morning sun and watch the lizard. Were those really push-ups, you wonder - like what I do at the gym? Why push-ups? Do lizards need their morning exercise? Was the lizard looking for food? Was it trying to scare me away, or was it checking for other lizards? Maybe it was watching for someone who might want a lizard for breakfast.

You look over at the tree trunk and notice the lizard on the ground, bobbing its head up and down very quickly. *Now, what?*

It's time to get your fishing gear. Walking to the lake, you keep thinking about what you saw. Why was that lizard doing push-ups—or was it? Why was it bobbing its head? Was it talking to me or some other lizard? People talk to one another. But lizards don't speak, do they? How do they let other lizards know what is happening if they don't talk?

Male Collared Lizard Displaying

What Lizards Do

Male Brown Anole Displaying Dewlap

The answer is that there are at least four ways that lizards tell others what's happening. Many lizards use more than one way to communicate with other lizards. Some do push-ups, some bob their head up and down, and some do both. Some reshape their whole body, and some use color to look big and fierce.

The green iguana, the Galapagos marine iguana, anoles, spiny lizards, desert iguanas, and side-blotched lizards do push-ups. They also bob their heads up and down and hold their bodies in specific ways. They puff up, flatten, and tilt their bodies to look as big as possible. Anolis lizards have a big, colorful flap under their chin. They tip their heads up and lower that flap while doing push-ups. Tegus and whiptail lizards bob their heads up and down, their way of saying, "Stay away from me" or "Hi, I like you!"

Indian Chameleon Changing Color

What Lizards Do

Some lizards change their color. Chameleons are champion color changers. Chameleons can change their color, and some change the patterns of their colors. They also hold their body in unique ways. A brightly colored chameleon with an arched back warns others to stay away. A plain-colored chameleon holding still is trying not to be seen.

Chameleon Using Color to Hide

Pexels

Push-ups, head bobbing, and color changes tell other lizards they need to move on or want to mate. Male lizards don't wish other male lizards to be around. Male lizards also use their signals to get female lizards to pay attention to them. Female lizards will move or change colors to say, "Hi, I like you," or "Go away." If a lizard ignores the message from another lizard, they will fight, and the loser leaves.

That lizard by your cabin did push-ups to tell you, "This is my tree trunk - go away."

You went away—that lizard is now looking for breakfast.

What's For Lunch?

You are back at your cabin. The fishing was good. You have a trout for lunch. As you sit on the edge of the porch munching your crunchy trout, you see a lizard jump off a fallen tree trunk and grab a beetle. The

Galapagos Land Iguana Eating Cactus Fruit

lizard munches the beetle much like you did your trout and climbs back onto the tree trunk.

"Hmm, the lizard must be having lunch too. I wonder if it is the same lizard I saw this morning. What do lizards eat besides insects?" you wonder.

That's a good question. What do lizards eat, and where do they find food? A lot of lizards primarily eat insects. Some eat birds' eggs, and

Marine Iguana Eating Seaweed

some eat small birds and mice. A few eat leaves, flowers, and fruit.

Green iguanas and Lesser Antillean iguanas climb trees and eat fruit and leaves. Spiny iguanas eat cactus fruit and plant leaves. Galapagos land iguanas eat cactus pads, fruit, insects, and eggs. Galapagos marine iguanas dive to the bottom of the ocean and munch the seaweed that

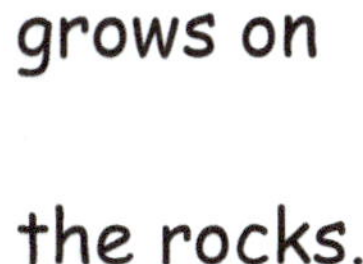

grows on the rocks.

Desert Iguana

Desert iguanas eat the leaves and flowers of desert plants.

Chuckwalla on a Rock

Chuckwallas live in rocky places in the Sonoran and Mojave

deserts. They munch on the plants that grow among the rocks.

North African Spiny-tailed Lizard

Spiny-tailed lizards live in deserts in Africa and Asia. Spiny-tailed lizards eat leaves, flowers, and fruit of desert plants.

Solomon Islands Skink

Solomon Islands skinks live in tropical forests and eat leaves, fruit, and new plant shoots. Those skinks climb trees and hang onto limbs and twigs with their tails while they eat fruit or leaves.

Side-blotched lizards in the southwest U.S. eat flowers while

Male Side-blotched Lizard

catching the insects on them.

Monitor lizards in Africa and Asia eat plants, insects, and other animals.

Nile Monitor

Asian Water Monitor

What Lizards Do

Blue-tongue skinks eat the snails that live in people's gardens.

Common Blue-tongued Skink

Australians like to have blue-tongued skinks in their yards.

Gila Monster

Gila monsters and Mexican beaded lizards are venomous and will eat baby mice and rats, but they prefer to eat birds' eggs. Those big, slow lizards find their food by poking around in holes, animal burrows, and under plants. They locate their food by smelling it with their tongue.

Mexican Beaded Lizard

Fast Food

"I'll Have a Number 6, Please."

Desert Spiny Lizard Watching for Food

I don't think there's a lizard McDonald's with a run-up window.

There's no corner market for them to go to. Lizards hunt for their

Red-headed Rock Agama Watching for Food and Doing a Pushup

lunch, or they eat things that are growing where they live.

Zebra-tailed Lizard Watching for Food

Hunting lizards have several ways to catch their prey. Some find a place, sit, and wait for their food to walk by. Spiny lizards in North America like to sit on tree limbs, tree trunks, or fences. When insects

What Lizards Do

Collared Lizard Watching for Food

run by, they grab and eat them. The guy you watched catch a beetle was a spiny lizard. The red-headed rock agama in Africa sits on rocks and watches for food. Zebra-tailed lizards in U.S. deserts sit and wait under the edges of bushes. Collared lizards sit on rocks or

New Mexico Whiptail Eating an Insect

under bushes and watch for smaller lizards or large insects. Collared lizards then dash out and catch their food.

Whiptail lizards in North America actively look for their food. A whiptail lizard searches like a bird. They use their noses, like birds use their beaks, to dig around in the leaves and stuff under bushes. They eat the insects that they find. If a whiptail finds an insect grub buried in the ground, it will dig it up and eat it. After eating a big grub, the whiptail will go into its burrow and take a nap. In South America, Tegus

Black and White Tegu Finding Food

hunt a lot like whiptails by searching on the forest floor. Tegus can smell with their tongues and find food that way.

Green Chameleon Watching for Food

Most chameleons live in trees or bushes. Chameleons can see in two directions at once. Each eye moves by itself, so they can even look backward. Their movable eyes help them find food and watch out for enemies. Chameleons sneak up on their food by walking slowly along the ground, tree limbs, or twigs. When they get close to an insect, they shoot their long, sticky tongues out of their mouths. Thwap! Their tongue hits the insect. They pull their tongue back and eat the insect stuck on the end.

Some lizards, like some people, are very picky about what they eat. Horned lizards in North America only eat ants. Different-sized horned lizards eat different-sized ants.

Texas horned lizards eat large, red harvester ants. Small round-

Texas Horned Lizard

tailed horned lizards eat small honey ants. Horned lizards will park near an ant nest and grab and munch down the ants that walk by.

Thorny devils in Australia, much like horned lizards, also eat ants.

What Lizards Do

Thorny Devil

Both thorny devils and horned lizards have a unique way to drink water. Thorny devils put their front feet into the water. The water moves up their legs in tiny grooves and into their mouths. Thorny devils also scoop wet sand onto their backs after it rains. The water from the sand moves along tiny grooves in their skin to their mouth. Horned lizards also drink by catching water on their skin. Horned lizards stand in the rain, and the raindrops that hit their skin move to their mouth through tiny grooves on their skin. Horned lizards also get water from wet sand using their skin.

Komodo Dragon

Komodo dragons are the largest lizards in the world. They can get ten feet long and weigh over 200 pounds. Komodo dragons eat goats, pigs, water buffalo, monkeys, birds, and other Komodo dragons. Dragons have venomous saliva and can kill their food if they have to, but they usually eat animals that are already dead. Komodo dragons can swallow whole animals, bones and all. Other dragons will also show up when one finds a dead animal. They may fight over the dead animal. After eating, Komodo dragons will sit in the sun and let their food digest.

Komodo Dragon With Food

Lizards and people both need rest after a meal. You have finished your lunch, so you decide it is time for a nap like a whiptail or a Komodo dragon. But you are not a Komodo dragon, so you throw your fish bones in the trash; you don't eat them.

Fenton R. Kay

Sand Swimmers, Water Walkers, and Gliders

Algodones Dunes at Dawn

On the first full day of your family's summer trip, you visited the Algodones Dunes State Park in southern California. You walked into the dunes to enjoy the desert sunrise and the cool of early morning. Suddenly, from right in front of you, a lizard popped out of the sand and dashed off. As you walked, you were looking very carefully at the ground because the park rangers had warned you about buried sidewinder rattlesnakes. You didn't see the lizard. It must have been buried in the sand. The lizard ran ahead of you for a few feet, stopped, wiggled, and disappeared into the sand. You just saw a fringe-toed lizard.

A lizard that lives on dunes. What an amazing animal. "*How do they do*

Mojave Fringe-toed Lizard in Sand

that—running across the sand, diving into it, and wriggling away?" you wondered.

Sand dunes are a different sort of place to live. Living on dunes is not easy. Sand lizards and other animals that live on dunes do some

Mojave Fringe-toed Lizard

strange things so that they can live in the sand. Wind blows sand across the land. When the wind stops, the sand falls to the ground—that is how dunes are made. As you walk along, you notice that the surface of the sand dunes is very loose and is hard to walk on. The sand shifts, slides, and sinks under your feet. Sand is very loose and moves about easily, like water. Did you try running? Wow, that was hard, wasn't it? So how does a fringe-toed lizard dash across the sand like

that? The answer is in the lizard's name. Scales on the lizard's toes poke out to the side. They make the lizard's foot bigger but not heavier. A broad, lightweight foot allows the lizard to run over the sand. The fringe scales also help the lizards wriggle under the sand to hide or sleep.

How neat! Lizards that run over the sand and wriggle into it. Are there other sand dune lizards like the one you saw? The answer is 'Yes'. Dumeril's fringe-toed lizards live in northern Africa and Arabia. Nambia, on the west coast of Africa, has a desert with huge dunes. Living there is a sand lizard called the shovel-snouted lizard. Shovel-snouted lizards and Dumeril's fringe-toed lizards look and act very much like North American fringe-toed lizards. They all have fringed toes and noses that work like shovels to help them move under the sand.

What would you think if I told you there were fish that live in dunes?

Sandfish Skink

"Aw, come on, fish don't live in sand," you would probably respond.

What Lizards Do

Yes, they do, sort of. There is a lizard called the sandfish in the Sahara Desert because it swims through the sand. Well, it doesn't

Juvenile Male Sandfish Skink

swim. It wriggles through the sand like a fish swim through the water. It doesn't use its feet or legs to move like fringe-toed lizards. The sandfish wiggles back and forth under the sand like a snake. The wriggling makes a wave along its body that pushes it through the sand. The sand-fish has smooth scales, and its nose is shovel-shaped, like fringe-toed lizards. It swims deep into the sand, where it hides or sleeps.

Fenton R. Kay

Dunes can be very hot or cold, which is part of why they are hard places to live. However, just a few inches under the sand's surface, it stays cool during the day and warm at night. That is why lizards and other dune-dwelling animals can survive, and some are good at sand-swimming.

How to Walk on the Water

If you think swimming in the sand is neat, how about walking on water? Nope, I'm not fooling you. Some lizards walk on water. A lizard called the basilisk lives in South and Central America. Basilisks are also called Jesus Christ lizards. Those water walkers live in trees and bushes along streams and rivers. When startled, they jump down and run across the water to escape.

Brown Basilisk

Plumed Basilisk Running on Water

Now, here's something else that is cool. Sand swimmers and water walkers have to move across places not designed for walking easily. Like fringe-toed lizards and sand-fish, basilisks have fringes of scales on

Green Basilisk Portrait

their toes. The scales make their feet extra wide and help to hold them up as they run across the water. If the other shore is too far away, they run as far as they can, dive into the water, and swim to safety.

It started to get warm, so you walked back to your campsite. As you went, you wondered about what other ways lizards might move around

Flying Dragon Displaying His 'Wings' and Dewlap

Here Come the Hang Gliders

Most lizards have four legs and can run and climb. Would you believe me if I told you that there are lizards that can fly? Well, not fly like birds, but glide through the air.

"No, I'm not kidding—just like hang-gliders!"

Flying lizards live in the forests of Southeast Asia, southern India, the Philippines, and Borneo. The lizard's scientific name, *Draco*, means

Flying Dragon Getting Ready to Launch

dragon. They live high in the forest's trees and glide from tree to tree. Flying dragon lizards have long ribs and lots of skin. When they jump from a tree, they spread their ribs out. The skin along their ribs

Flying Dragon *Flying*

makes a wing that lets them glide for long distances. Flying dragon lizards land on the trunk of another tree, scurry up, and then jump off again. Just imagine what it would be like to walk through the forest and look up to see a lizard flying by.

Draco's are not the only lizards that glide. They share the forest with a flying or gliding gecko. Geckoes are active at night, and *Draco*'s by day. Flying geckos also live high in the forest trees. When they jump off a tree, they don't open their ribs like Draco. Gliding geckoes have

Kaheng Kratchan Flying Gecko

skin flaps along their heads, bodies, and tails. They also have webs between their toes. The folds of skin and webs make them very flat, allowing them to paraglide from one tree to another. *So, I'll bet you think there can't be any more flying lizards—that's just too strange.*

Nicobar Gliding Gecko Camouflaged on Tree Trunk

In Africa, however, a small lizard lives in the forest in tall trees. Guess what? That little lizard is a blue-tailed gliding skink. Blue-tailed gliding skinks are very flat. Some of their toes have a web between them, which makes their feet extra wide, and their bones are very light. Like Draco, flying geckos, they jump from one tree and glide to the next tree.

Blue-tailed Gliding Skink

Blue-Tailed Gliding Skink with Body Flattened for Gliding

What Lizards Do

As you return to your campsite, you decide to get online and look up flying, water-walking, and sand-swimming lizards. When you start, your mom is amazed. You really do want to find more neat things about lizards.

Keeping Your Home, Using Color, and Making Babies

Nearly all lizards, especially males, are territorial. That is, they have a patch, and they defend it against all comers. Territories are important to lizards for much the same reasons that 'homes' are important to people. A lizard's territory—its patch—provides everything needed for life: food, a safe place to hang out, and a place to make babies so there will be lizards in the future. A male lizard needs a patch that is his alone, and it will be interesting to the neighborhood girl lizards.

Hey, This Is My Place

Male Rock Agama Warning Other Males to Stay Away

What Lizards Do

Lizards have a lot of different ways to declare that 'this is my patch.' As we have seen before, many lizards do pushups or bob their heads to warn off other lizards. Some lizards will make themselves look big and fierce by flattening their sides or arching themselves into an upside-down U. Other lizards, especially ones that are active at night, use pheromones—stinky chemicals that tell other lizards of the same kind that this is my patch. Sometimes, they will rub the part of themselves

Zebra-tailed Lizard Male Displaying Orange and Yellow on His Side

that produces the chemicals on rocks, twigs, plant stems, and other places around the edge of their patch. A sort of stink fence. Other lizards leave their chemicals on their poop and put poop piles in places that mark the boundary of their patch. All of these are ways lizards announce to the lizard world that they own that patch, and you'd best not be coming in.

Anole Male Displaying Dewlap

Anoles are usually plain brown or green, sometimes with stripes or spots. Male anoles can display a big flap of skin under their chins. The flap is called a dewlap and is usually some bright color, different than the lizard's body color. The lizards raise themselves on their front legs and lower the dewlap. Sometimes, they raise and lower it several times, like waving a flag. Male anoles use the dewlap to signal to other males that this is their patch and to stay away. They also display their beautiful dewlap to impress the girls.

Color Changes

In many kinds of lizards, only the males change colors. In chameleons, both males and females can and do change color. Color change in chameleons is both a protection and a way to keep others away. If a chameleon senses danger, they will match their color to wherever they are. Sometimes, chameleons use bright colors to say, 'I'm angry, stay away' or 'I'm looking for a mate—aren't I beautiful.'

Madagascar Chameleon Changing Color

What Lizards Do

Many lizards that do pushups and head bobs also have bright patches of color—often blue—on their throats and along the sides of their bellies. During the mating season, they display their blue throats during their pushup routines. If another male comes around, they will flatten themselves and tilt their bright blue sides into clear view of the other lizard.

Blue Crested Indo-China Forest Lizard on a Tree Trunk

Chuckwalla Warming Up on a Rock

Many sun-warmer lizards, like the fence lizards, desert spiny lizards, and chuckwallas, turn very dark, almost black, when they are cold. When they move into the sun, the black color helps them warm faster. Once they get to whatever temperature they need, they resume their normal color, often some shade of brown. Lighter colors warm more

Desert Spiny Lizard Warming Up

slowly, which is important when they are moving around seeking food or a mate.

Green Madagascar Day Gecko on Tree Limb

Many of the ground-warmers, skinks, and geckos, for instance, don't change color. Because their environment is nearly always warm, turning dark doesn't help. Big, green tropical iguanas seldom change color. Their green matches the trees where they live.

Laying Eggs

Birds always lay eggs. Most lizards lay eggs, but some don't.

Bougainville's Skink Lays Eggs and Has Live Young

Bougainville's skink both lays eggs and has live young. Lizards lay eggs in many sorts of places. House geckos need a dark, moist place that

Female European Green Lizard Laying Eggs

stays warm. They lay their eggs in old piles of leaves or flower pots. Many lizards lay their eggs in warm, sandy soil. The female will dig a hole, lay her eggs, and cover them with sand or soil.

The baby lizards will hatch underground, dig themselves out, and start looking for food.

Most lizards do not take care of their young. After laying her eggs, the female lizard finds food and avoids becoming someone else's food.

Female Eastern Glass Lizard Protecting Her Eggs

Male lizards have nothing to do with the young. Female glass lizards stay with their eggs until the young hatch.

More than one female gecko may lay their eggs in the same place, and some gecko females use the same egg-laying site year after year. In some skinks, several females will lay their eggs in the same place—

Eggs of Several Skinks in the Sand

under a slab of rock or a fallen tree. Some male skinks will guard the egg-laying sites of females. They do that so they will have a better chance of finding a mate.

Other lizards, like chameleons, don't lay eggs but produce live young. The female lizard finds a place to hide when ready to have her babies. The female will make the young lizards, still wrapped in the sacs—

membranes—they would have if they were in eggs. The young lizards

Mexican Beaded Lizard Hatching

will break out of the sacs and leave the birthing site. The female gives birth to the young and then returns to her regular life.

Baby lizards are on their own when they are born or hatch. Baby lizards look like very small adults. They usually eat the same things as adult lizards, but in smaller-sized bites.

Little Lizards

For instance, adult house geckos eat adult cockroaches. Baby house geckos eat baby cockroaches.

Juvenile Green Iguana

Many baby lizards are colored differently from adults. In many lizards, the young have brightly colored tails. Those lizard's tails break off easily, and the wiggling bright tail confuses the critter trying to eat the lizard. New Mexico whiptail young have bright blue-green tails, and the whole tail is colored. The adults' tails are often paler blue, with only the tip colored. Other lizards with bright-colored tails as young have plain lizard-colored tails as adults.

Juvenile New Mexico Whiptail

Farewell Lizards

Lizards are creatures of wonder. They are endlessly fascinating, especially if you think about the fact that they have been scuttling, climbing, running, swimming, and gliding around this planet far longer than humans. Lizards were first recognizable as lizards when the dinosaurs still ruled our planet. Today, lizards are the second most abundant terrestrial vertebrates—animals with a backbone—that live on land.

The most common animals with backbones in the world are birds. There are about 10,000 species (kinds) of birds. There are currently known 6,399 kinds (species) of mammals. People are mammals - we have a backbone and hair, are self-warmers, and produce milk. Lizards come in at 4,675 species. The snake count is 3,400 species. The most abundant land critters are insects; there are more than five million of them, and new species are being discovered daily.

Fenton R. Kay

For the curious naturalist, there is an almost unlimited supply of amazing and fascinating critters in the world, most of them just waiting to be studied and understood. There is still much to be learned about what lizards do, why they do it, when, and where they do it. Next time you are sitting in your yard and a fence lizard climbs up the wall, sits in the sun, and does push-ups and head bobs at you, close your eyes and imagine what that lizard's distant cousin may be doing somewhere across the world.

Names and Homes of Lizards

Common Name	Scientific Name	Home
Mediterranean House Gecko	*Hemidactylus turcicus*	Cosmopolitan
Duméril's Fringe-toed Lizard	*Acanthodactylus dumerilii*	North Africa
Marine Iguana	*Amblyrhynchus cristatus*	Galapagos Islands
Indian Chameleon	*Chamaeleo zeylanicus*	Sri Lanka, India, Pakistan
Galapagos Land Iguana	*Conolophus subcristatus*	Galapagos Islands
Collared Lizard	*Crotaphytus collaris*	Mojave, Sonoran, and Chihuahuan Deserts of Mexico and the U.S.

Fenton R. Kay

Kaheng Kratchan Flying Gecko	*Gekko kaengkrachanense*	Asia
Green Iguana	*Iguana iguana*	Mexico. Central & South America
Texas Horned Lizard	*Phrynosoma cornutum*	Chihuahua Desert of Mexico and the U.S.
Blue-tailed Skink	*Plestiodon elegans*	China, Vietnam, Taiwan, Japan
Five-lined Skink	*Plestiodon fasciatus*	United States, Canada
Wall Lizard	*Podarcis muralis*	Europe
Black and White Tegu	*Salvator merianae*	South America
North African Spiny-tailed Lizard	*Uromastyx acanthanura*	North Africa

What Lizards Do

Asian Water Monitor	*Varanus salvator*	Asia
Blue-tailed Gliding Skink	*Holaspis guentheri*	Africa
Bougainville's Skink	*Lerista bougainvillii*	Australia, Tasmania
Brown Anole	*Anolis sagrei*	Caribbean, U.S., Mexico
Brown Basilisk	*Basiliscus vittatus*	Mexico, Central & South America
Chuckwalla	*Sauromalus obesus*	Mojave and Sonoran Desert of Mexico and the U.S.
Common Blue-tongued Skink	Scincoides scincoides	Australia
Desert Iguana	*Dipsosaurus dorsalis*	Mojave and Sonoran Desert of Mexico and the U.S.
Desert Spiny Lizard	*Sceloporus magister*	Western United States, Mexico

Fenton R. Kay

Eastern Fence Lizard	*Sceloporus undulatus*	Eastern and Western United States
Eastern Glass Lizard	*Ophisaurus ventralis*	Southeastern United States
European Green Lizard	*Lacerta viridis*	Europe, England
Flying Dragon	*Draco volans*	Asia
Gila Monster	*Heloderma suspectum*	Mojave and Sonoran Desert of Mexico and the U.S.
Green Chameleon	*Furciver viridis*	Madagascar
Green Day Gecko	*Phelsuma madagascariensis*	Madagascar
Komodo Dragon	*Varanus komodoensis*	Komodo Islands

What Lizards Do

Mexican Beaded Lizard	*Heloderma horridum*	Mexico, Central America
Mojave Fringe-toed Lizard	*Uma scoparia*	Sonoran Desert Of Mexico and California
New Mexico Whiptail	*Aspidoscelus neomexicanus*	Chihuahua Desert of Mexico and the U.S.
Nicobar Gliding Gecko	*Gekko nicobarensis*	Nicobar Islands
Nile Monitor	*Varanus niloticus*	East Africa
Plumed Green Basilisk	*Basiliscus plumifrons*	Mexico, Central & South America
Red-headed Rock Agama	*Agama agama*	Africa

Fenton R. Kay

Sandfish Skink	*Scincus scincus*	North Africa, Arabia
Side-blotched Lizard	*Uta stansburiana*	Western United States and Mexico
Solomon Islands Skink	*Corucia zebrata*	Soloman Islands
Thorny Devil	*Moloch horridus*	Australia
Western Spiny Lizard	*Sceloporus occidentalis*	Western United States and Mexico
Zebra-tailed Lizard	*Callisaurus draconoides*	Mojave and Sonoran Desert of Mexico and the U.S.
Blue Crested Indo China Forest Lizard	Calotes mystaceus	Bangladesh, Cambodia, China, India, Laos, Myanmar, Thailand, Vietnam

ISBN: 13-979-8-9945355-2-3

Revised 1st Printing

Published by KayLibros Press

Acknowledgments

I am particularly indebted to Dr. James E. Deacon, under whose guidance I did my M.S. Thesis on lizard biology at a warm spring complex in the southern end of Death Valley. I also learned much of what I know about lizard behavior from Dr. Charles Carpenter, who first described and catalogued the species-specific head bobbing and pushup behavior of lizards. For the rest of my knowledge, I have gained from years of watching, reading, and thinking about lizards and what they do.

My wife, Dr. Carol Kay, read early drafts of the book. Her comments and criticisms have helped to make a better book. Several of my colleagues in the Las Cruces Writers Group also read and commented on an early draft of the book. I want to thank Desiree Bustamente and Margaret Loring, both teachers, and Frank Lopez for their willingness to read and comment on this work.

The pictures in the book were obtained from various online sources. All were obtained through the purchase of publication rights and are attributed in each picture. Vector images are all from 123RF. The cover photo is from Pexels. I use Grammarly ™ as a line-editing tool, but the writing is solely mine.

About the Author

Fenton R. Kay, Ph.D., is a retired biologist with fifty-plus years of experience working out in nature, wondering about the meaning of what he saw. Fenton wrote a master's thesis about the activity patterns of lizards at Saratoga Springs, Death Valley National Monument. He has published several scientific papers about lizards, snakes, and frogs. Fenton studied briefly under Dr. Charles Carpenter, the reptile behaviorist who described the details of lizard head bobs and pushups.

Besides this book, Fenton has published thirteen others, mostly mysteries. Some of his mysteries reflect places he has been and things he has done as a biologist. Others represent flights of fancy or conjectures on how things might be in the future or might have been in the past.

Fenton lives in a passive solar home designed by his wife in the northern Chihuahuan Desert. They have eight cats, three dogs, a red-eared slider turtle with its own pond, and two aquaria. A large population of house geckos, whiptail, side-blotched, and fence lizards are also resident on the property. The occasional rattlesnake, gopher snake, kingsnake, or red racer turns up. Coyotes hop the wall into the backyard. Skunks dig under the wire fence, looking for dog food. Pack rats get into the backyard shed, chew up everything, and eat his plants. A quail nested in an old, upside-down chair in the carport and produced a clutch of little brown chicks. A female hummingbird owns his backyard pergola during the spring and summer.

www.ingramcontent.com/pod-product-compliance
Lightning Source LLC
LaVergne TN
LVHW081634120826
845149LV00025B/1907
* 9 7 9 8 9 9 4 5 3 5 5 2 3 *